RAND NATIONAL DEFENSE RESEARCH INSTITUTE

United States Service Academy Admissions

Selecting for Success at the Military Academy/West Point and as an Officer

Lawrence M. Hanser, Mustafa Oguz

Prepared for the Director of Accession Policy, Office of the Undersecretary of Defense, Personnel and Readiness

For more information on this publication, visit www.rand.org/t/rr723

Library of Congress Cataloging-in-Publication Data
ISBN: 978-0-8330-8874-1

Published by the RAND Corporation, Santa Monica, Calif.

Cover Photo by Eric S. Bartelt/Pointer View via Flickr.

Preface

This report answers two key questions by examining the relationship between pre-admission information on applicants to the United States Military Academy at West Point (USMA) and two subsequent outcomes. First, do applicants with better results on USMA's "whole candidate score" (WCS), a major element in admissions decisions, have a greater probability of graduating? Second, do applicants with higher WCSs have a greater probability of remaining in the U.S. Army and being promoted to lieutenant colonel? These outcomes are important because when a cadet enters but does not graduate, he or she fills a class seat that could have been filled by someone else who might have graduated, and the cost cannot be recouped. The high cost of educating each cadet is an investment the Army makes in expectation of years of productive service to the nation beyond graduation. The relationship between application scores and outcomes are analyzed with logistic regression and a boosted logistic regression. The application scores found to be statistically significant are reported as having an influence on these outcomes.

This report is intended for Department of Defense policymakers who have responsibility for oversight of the U.S. military academies. It may also be of interest to U.S. Army policymakers and USMA officials. A companion report (Hardison, Burkhauser, and Hanser, forthcoming) focuses on admissions to the U.S. Air Force Academy.

This research was sponsored by the Director of Accession Policy, Office of the Undersecretary of Defense, (Personnel and Readiness) and conducted within the Forces and Resources Policy Center of the RAND National Defense Research Institute, a federally funded research and development center sponsored by the Office of the Secretary of Defense, the Joint Staff, the Unified Combatant Commands, the Navy, the Marine Corps, the defense agencies, and the defense Intelligence Community.

For more information on the RAND Forces and Resources Policy Center, see http://www.rand.org/nsrd/ndri/centers/frp.html or contact the director (contact information is provided on the web page).

Contents

Figures and Tables

Figures

Tables

Summary

This report analyzes data from United States Military Academy at West Point (USMA) applicants for graduation years 1992 to 2001 to identify whether the application scores that USMA uses as admission criteria can predict success both in the USMA and as an officer. We defined success at the USMA as a binary variable that equals 1 if an admitted candidate has graduated from the USMA and 0 if he or she did not graduate. We used two variables to define later success as an officer: early promotion to an O-4 pay grade and promotion to O-5.[1] The application scores we used to predict these three outcomes are the whole candidate score (WCS), which is calculated by the USMA, and the application scores that the USMA uses to calculate the WCS: the academic composite score, or college entrance examination rank (CEER), which factors in Scholastic Aptitude Test (SAT) or ACT scores and the high school rank convert score (HSRCS); candidate fitness assessment (CFA); faculty appraisal score (FAS); athletic activities score (AAS); and extracurricular activities score (EAS).

We used four models in our analysis. The purpose of examining several models is to assess whether there is a more predictive combination of applicant data than the current WCS.

The first model is a logistic model that only uses the WCS calculated by the USMA. Our goal was to understand if the WCS, an

[1] Early promotion to O-4 is defined as being promoted to O-4 before ten years upon receiving the first commission. According to Department of Defense Instruction 1320.13 (July 22, 2009), the desired active-duty list promotion timing to O-4 is ten years (plus or minus 1 year) and to O-5 is 16 years (plus or minus 1 year).

important part of the admission decision, is associated with success as a student and as an officer.

The first variable of the second model is the CEER, a component of the WCS that is calculated by the USMA using SAT/ACT scores and HSRCSs. Because it is another score calculated by the USMA, we tested if and how it is related to important outcomes. The second model also includes CFAs, FASs, AASs, and EASs. This model helps us understand if athletic activities, extracurricular activities, and high school faculty appraisals are related with success as a student and as an officer.

The third model breaks the CEER down to its two components (the SAT/ACT score and HSRCS) and includes CFAs, FASs, AASs, and EASs. The purpose of the third model is to see if SAT/ACT scores and high school ranks are separately related with better outcomes—and if so, how.

We also included a fourth model that uses boosted logistic regression on the same variables in the third model. Boosted logistic regression has the advantage of aggregating predictions from many weak regression tree classifiers and outperforming a single logistic model in predicting outcomes. The boosted logistic model also explores the increase in predictability that results from interactions between predictors. The purpose of the boosted regression model is to estimate the best possible predictive capacity of the admission variables against which to compare the results of the simple logistic regression models.

Our analysis led to several observations. The first is that the application scores used by the USMA do a good job of predicting graduation outcomes. In the first model, the WCS is statistically significantly associated with higher rates of graduation. As the WCS of a candidate increases, so does the probability of that candidate's graduation. In the second model, the CEER, CFA, AAS, EAS, and FAS are statistically significant predictors of the probability of graduation. The third model found that the SAT/ACT score is significant but that the HSRCS is not associated with the probability of graduation. The boosted logistic regression model suggested that the converted SAT/ACT score, FAS, and HSRCS have the highest influence on graduation outcomes. The difference regarding the significance of the HSRCS between the third and fourth models indicates that there is an interaction effect that the simple logistic model does not capture.

Next, we tested whether the application scores are related to early promotion to O-4 and promotion to O-5. We used the same four models, but because promotions might be related to the cohort year of candidates, we added cohort years as additional explanatory variables to each model. We observed no relationship between application scores and the probability of early promotion to O-4, but found that having a graduation year of 1998, 2000, or 2001 had the largest positive effect on the chances of early promotion to O-4, according to all four models.

Finally, we observed that the WCS is a statistically significant predictor of promotion to O-5, with higher WCSs associated with higher probabilities of promotion. The academic composite score of the USMA, the CFA, the EAS, and the FAS are associated with better chances of promotion to O-5. The AAS did not influence the promotion outcome. In the third model, higher SAT/ACT scores, HSRCSs, CFAs, and FASs are associated with higher probabilities of promotion to O-5. The AAS and EAS are not related to outcomes regarding promotion to O-5. The boosted logistic regression model found that the graduation year of the candidate has the biggest influence on the O-5 promotion outcome. Graduation years 1997, 1998, and 1999 have the largest negative effect on probability of promotion to O-5, which is expected because candidates from these later years had less time to be promoted than candidates from earlier years.

Thus, the results from the four models suggest that the use of WCSs as the primary basis for admission decisions at West Point is warranted.

Acknowledgments

The authors wish to acknowledge the originators of this study within the Office of the Secretary of Defense—the late Mr. Bill Carr, then–Deputy Under Secretary for Military Personnel Policy, and Dr. Curtis Gilroy, then–Director of Accession Policy within the Office of the Secretary of Defense (Personnel and Readiness). Mr. Jeffrey Mayo, the Director of Accession Policy who followed Dr. Gilroy, and our points of contact, Mr. John Jessup, Lieutenant Colonel Paul Nosek, Lieutenant Colonel Jason Knight, and Mr. Dennis Drogo, proved instrumental in shepherding this study over the course of time and bringing it to a successful conclusion. We wish also to thank the Director of Admissions, Colonel Deborah J. McDonald, and the members of the admissions office at the United States Military Academy at West Point who were so gracious in providing the data on which our analyses are based. Mr. Scott Seggerman and others at the Defense Manpower Data Center also provided access to data, without which our analyses could not have been completed. First Lieutenant Jeremy Didier, then a graduate student at the RAND Pardee Graduate School, was largely responsible for the early cataloging and organizing of the data used in this study. We are grateful to Dr. Kata Mihaly and Dr. Gilroy for their thoughtful reviews of this report. The authors are responsible for any remaining errors of omission or commission.

Abbreviations

AAS	athletic activities score
CEER	college entrance examination rank
CFA	candidate fitness assessment
CLS	community leader score
DMDC	Defense Manpower Data Center
EAS	extracurricular activities score
FAS	faculty appraisal score
FYGPA	first-year grade-point average
GAO	General Accounting Office
GPA	grade-point average
HSGPA	high school grade-point average
HSRCS	high school rank convert score
MPS	military performance score
OCS	Officer Candidate School
ROC	receiver operating characteristic
ROTC	Reserve Officer Training Corps
SAT	Scholastic Aptitude Test
USMA	United States Military Academy at West Point
WCS	whole candidate score

CHAPTER ONE

Introduction

The RAND National Defense Research Institute was asked by the Office of the Undersecretary of Defense (Personnel and Readiness) to examine admission standards at the military academies to see whether the academies were enrolling individuals who would not only graduate from the academies but would also be successful officers in their respective services. This report focuses on admissions to the United States Military Academy at West Point (USMA). A companion report (Hardison, Burkhauser, and Hanser, forthcoming) focuses on admissions to the U.S. Air Force Academy. The U.S. Naval Academy declined to participate.

The USMA has a long history, being first established by Congress in 1802. After his appointment as superintendent in 1817, Sylvanus Thayer "made West Point America's national engineering school" (Smithsonian Institution, undated). It remained essentially an engineering school until the mid-20th century, when Superintendent Maxwell Taylor added humanities and social studies classes to the curriculum. A few years later, Garrison Davidson changed the curriculum from one where all cadets followed the same schedule of courses to a curriculum that allowed elective courses to be taken (Smithsonian Institution, undated). USMA strives to balance its mission of preparing and commissioning Army officers with its continuing reputation as an undergraduate engineering school. Its mission statement in the *2011–2012 West Point Catalog* (USMA, undated-a) reads:

> To educate, train, and inspire the Corps of Cadets so that each graduate is a commissioned leader of character committed to the values of Duty, Honor, Country and prepared for a career of pro-

fessional excellence and service to the nation as an officer in the United States Army.

In its admissions process, current as of 2013, USMA calculates and uses in its admissions decisions a "whole candidate score" (WCS) for each applicant, which consists of a weighted combination of standardized test scores, high school rank (corrected for class size), number and level of extracurricular and athletic activities, physical fitness, and a series of ratings from one of an applicant's high school teachers. Applicants must also complete three short essays and be nominated by a member of Congress or other official legally qualified to make nominations to the military academies. Successful applicants must be medically qualified or receive a medical waiver.

In addition, another route to admission to West Point is through the USMA Preparatory School. Applicants to West Point who are not deemed qualified to meet admission standards may be offered admission to the Preparatory School and after a year may reapply to West Point. The prep school is an important source of diverse cadets for West Point. "The importance of the Prep School as a stepping-stone to West Point is evident. Since 1951, West Point Prep graduates have comprised 11 percent of the Corps of Cadets, yet they have held 25 percent of the senior leadership positions of the Corps" (USMA, undated-b).[1]

Desired Outcomes of the Admissions Process

USMA endeavors to admit students who will be successful both as students and as Army officers. In a memorandum to Admissions Department staff, the Director of Admissions wrote:[2]

[1] For a discussion of the role of preparatory schools and diversity, see U.S. Department of Defense (1999).

[2] The six class composition goals target the admission of "Scholars, Leaders, Athletes, Minorities, Women, Soldiers" (McDonald, 2008). These goals are related to what has come to be called the "whole person" or "holistic" review process in college admissions, which strives to account not only for the strength of each applicant's academic background but also a diversity of life experiences.

> Our primary mission is to make all possible efforts to inspire and enroll a diverse, high quality class of cadet candidates who are motivated toward *completion of West Point* and a *military career*. We will take all efforts to meet *class composition goals* without sacrificing quality. (McDonald, 2008; emphasis added.)

In this study, we focused on two desired outcomes from the admissions process: graduation from USMA and success as an Army officer. Graduation from USMA is one of three major sources of Army officer commissions. The other two avenues to officer commissioning are the Reserve Officer Training Corps (ROTC) program, which exists at many colleges and universities across the country, and Officer Candidate School (OCS), which provides opportunities for those who are already college graduates to be commissioned after successfully completing a short period of training.

Graduation from USMA is an important marker for the admissions process because it is the first hurdle that USMA cadets must pass on their way to becoming an Army officer, and each class seat that goes to a cadet who fails to graduate represents a lost and expensive opportunity that could have been given to a successful candidate. According to a General Accounting Office (GAO, now Government Accountability Office) report, the cost of producing an officer in the class of 1995 was $277,000 at the Military Academy (GAO, 1997). While not all successful applicants remain at USMA to graduate, the average graduation rate for the ten graduation years we examined in this study (1992–2001) was 76 percent, meaning 3,001 individuals were admitted who did not graduate.

Success as an Army officer is more difficult to measure. We used two metrics to assess this. The first was early promotion to rank O-4 (major) relative to the average time it takes an officer to reach that rank. We do not have an indication of below-the-zone promotion for any ranks, so we used ten years as the demarcation between early and average time to promotion to rank O-4. The second outcome we used to indicate success was achieving promotion to O-5 (lieutenant colonel).

Research on the relationship between college admission scores and college success indicates that there is a positive correlation among Scholastic Aptitude Test (SAT)/ACT scores, high school grade-point averages (HSGPAs), and college success. Geiser and Studley (2002) found that SAT scores are positively correlated with first-year grade-point averages (FYGPAs), using University of California Corporate Student System data on FYGPAs, SAT scores, and HSGPAs of first-time students entering freshman year between 1996 and 1999 (N = 77,893). SAT scores explain 16.2 percent of the variation in the FYGPA, and the HSGPA explains 15.4 percent of the variation in the FYGPA on average between 1996 and 1999. Together, the two scores explain 22.3 percent of the variance.[3] Geiser and Santelices (2007) found that SAT scores and HSGPAs are also positively associated with graduation after controlling for parents' education, family income, and school academic performance index ranking of students' high schools. We do not observe the last three variables in our dataset. We also do not have information on the HSGPAs of the candidates, but we have used high school ranking convert scores (HSRCSs) in our analyses. Niu and Tienda (2012) analyze data from the Texas Higher Education Opportunity Project and report that high school rank and test scores are significantly correlated with four-year grade-point averages (GPAs).[4] DeAngelo et al. (2011) report that SAT scores and HSGPAs are significantly associated with four-, five-, and six-year graduation outcomes after controlling for variables such as student family education level, student family income, and the allocation of the student's time in the last year of high school.

Study Approach

Complete application data on all applicants to the graduation classes of 1992 to 2001 were made available to us by West Point. However, we did not obtain detailed outcomes on the performance of cadets while at West Point, such as GPAs or graduation order of merit. To obtain outcome

[3] HSGPAs and SAT scores combined cannot explain as much as their individual totals because of the correlation between the two scores.

[4] Niu and Tienda (2012) convert ACT scores to SAT scores in their analysis and use the SAT scores as test results. In this study, we converted the ACT scores to SAT scores.

variables for our analyses, we also obtained career personnel information from the Defense Manpower Data Center (DMDC) on all applicants for whom West Point had provided data to us and merged that into our analysis dataset. Although many West Point applicants were admitted to and commissioned by West Point, a substantial number of applicants who did not enroll in West Point received their commissions from the Naval or Air Force Academy or the ROTC.

To evaluate the predictive validity of applicant information, we created three longitudinal datasets. Because many individuals apply for admission several times over, we used the applicant record from each applicant's latest application for each dataset. In order to examine the predictability of graduation from West Point from applicant data, the first included all applicants who had been admitted to West Point. In order to examine the predictability of early promotion to the rank of major, the second included all applicants with complete applicant data who had been promoted to the rank of major (O-4). In order to examine the predictability of promotion to the rank of lieutenant colonel (O-5) from applicant data, the third included all applicants with complete applicant data who had been promoted to the rank of major (O-4).[5] We conducted a variety of statistical regression analyses to explore the following research questions:

1. Is a WCS predictive of graduation from the USMA? Are components of a WCS predictive of graduation from the USMA?
2. Is a WCS predictive of promotion to O-4 within ten years of receiving the first commission? Are WCS components predictive of promotion to O-4 within ten years of receiving the first commission?
3. Is a WCS predictive of promotion to O-5? Are WCS components predictive of promotion to O-5?

We used three logistic regression models and a boosted logistic regression model to evaluate if and how the WCS and its components

[5] 2000 and 2001 cohorts are excluded from the O-5 promotion analysis because no candidate was promoted to O-5 as of 2012.

are related with the three outcomes. The first model includes only WCSs for the graduation analysis, an important input for the final admission decision, to assess its relationship with the outcomes. The second model includes the academic composite score, or college entrance examination rank (CEER)—which is calculated by the USMA from the SAT/ACT scores and high school rankings of candidates—along with other WCS components.[6] The third model breaks CEER down to SAT/ACT scores and high school rankings and also includes other WCS components. The different variables in logistic regression models aim to find out if the scores calculated by the USMA (WCSs and CEERs) are related with the three outcomes. The boosted regression model uses the variables in the third model to assess the relative influence of the scores on the outcomes. The boosted regression model iteratively fits regression trees to the error from a logistic model, which produces a better fit model.

Organization of the Report

Chapter Two provides a snapshot of application and acceptance rates at USMA for the graduating classes of 1992 to 2001 and describes the components of the application process at USMA. Chapter Three describes the data used in our analyses of the graduation outcome, the regression models we applied, and results of those models. Chapter Four describes the data used in our analyses of the promotion outcomes, the regression models we applied, and results of those models. Chapter Five presents our conclusions. Appendix A details the construction of our analysis samples. Appendix B provides a detailed comparison of the outcomes of the different regression models for predicting promotion to O-5. Appendix C displays the correlations among the variables used in our analyses. Appendix D describes the proper interpretation of the coefficients from our logistic regression models.

---

6 All WCS components are described in the next chapter. These components are athletic activities score (AAS), extracurricular activities score (EAS), faculty appraisal score (FAS), and candidate fitness assessment (CFA).

CHAPTER TWO

Applying to West Point

Overview

Applying to West Point is a long process that starts at the beginning of the candidate's junior year in high school. Candidates must be U.S. citizens (with the exception of a very small number of international admissions); they must not be married, pregnant, or have a legal obligation to support a child; they must have no dependents and must be at least 17 years of age but not yet 23 years of age.

Data suggest that, on average, approximately 12,500 applications were submitted to West Point for each graduation year between 1992 and 2001. Approximately 13 percent of these applicants received an offer, approximately 10 percent were admitted to West Point (see Table 2.1), and approximately 74 percent of those who were offered admission actually enrolled.

West Point emphasizes and scores three important qualities for each candidate: academic ability, leadership potential, and physical aptitude. Scores on these traits are combined to make up the WCS. An academic ability score is calculated by combining a candidate's ACT or SAT scores (whichever is the highest in terms of percentile standing) and high school rank,[1] making up 60 percent of the WCS (see

[1] "The minimum SAT scores for the academies are 500 verbal and 500 math. The average SAT scores at the academies are 540–620 verbal and 630–710 math. The minimum ACT scores for the academies are 21 English, 19 Social Studies, 24 Mathematics, and 24 Natural Science. The average ACT scores are 23–27 English; 24–29 Social Studies; 27–32 Mathematics; and 28–32 Natural Science (minimum and average scores are slightly higher for the

Table 2.1
West Point Applications and Acceptance Rates

Graduation Year of the Application	Applications	Received Offer	Accepted Offer	Percentage of Applicants Who Received and Accepted Offer
1992	13,860	1,761	1,311	9
1993	12,121	1,787	1,341	11
1994	12,139	1,777	1,322	11
1995	11,661	1,635	1,248	11
1996	13,127	1,655	1,189	9
1997	12,700	1,599	1,211	10
1998	12,521	1,615	1,147	9
1999	11,812	1,618	1,187	10
2000	12,262	1,561	1,186	10
2001	12,698	1,598	1,192	9
Total	124,901	16,606	12,334	10

Figure 2.1). Leadership potential is measured by the community leader score (CLS), calculated by taking the arithmetic average of the AAS, EAS, and FAS. CLS makes up 30 percent of the WCS. The final component is the physical aptitude exam score, which makes up the last 10 percent of a candidate's WCS.

In addition to the WCS, candidates are evaluated on three essays, and every candidate needs a nomination from an eligible sponsor for consideration. Next, we describe each application component.

Components of the Application

SAT and ACT Scores

USMA uses each candidate's best SAT or ACT score with regard to percentile standing when calculating a candidate's academic score. SAT takers submit SAT verbal and SAT math scores, and ACT takers submit

Naval Academy). Virtually all cadets are from the top 25 percent of their high school class." (White House, undated)

Figure 2.1
Components of the Whole Candidate Score

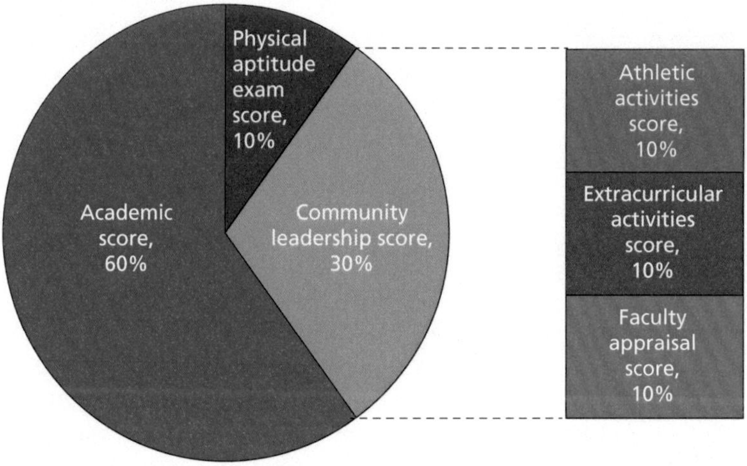

RAND *RR723-2.1*

ACT English, ACT Math, ACT Natural Sciences (Science Reasoning after 1989), and ACT Social Studies (Reading after 1989) scores.

High School Rank Convert Score

USMA calculates an HSRCS by using the candidate's rank in his or her high school and the size of the candidate's class. Candidates who have higher ranks in larger classes have a higher HSRCS.

Athletic Activities

Candidates' athletic participation and successes during their high school years determine their AAS. This score ranges between 200 and 800. Points are assigned as described in Table 2.2.

Extracurricular Activities

A candidate's participation in school activities outside the school curriculum determines his or her EAS. Points are assigned as described in Table 2.3.

Table 2.2
Athletic Activity Levels and Points

Athletic Activity Level	Points
An outstanding athlete (All-American, 1st team All-Area selection in baseball/softball, basketball, or football) and athletic rating of either 1 or 2 in the sport in which honors are received or a CFA score greater than 650.	800
1st team All-Area selection in a single sport (other than baseball/softball, basketball, or football). Captain of baseball/softball, basketball or football team. Team captain in two or more sports (other than baseball/softball, basketball, or football, for class size over 100).	700
Captain of team (other than baseball/softball, basketball, or football). Varsity letter in baseball/softball, basketball or football. Varsity letter in two or more sports (other than baseball/softball, basketball, or football).	600
Varsity letter in a single sport (other than baseball/softball, basketball, or football).	500
Participation in varsity sport without a letter.	400
Participation in junior-varsity and other team sports (excluding intramurals).	300
Candidates with no participation and no evidence of interest in sports.	200

Table 2.3
Extracurricular Activity Levels and Points

Extracurricular Activities Level	Points
An outstanding young person with quadruple participation or honors and awards on selected extracurricular activities (each worth 600 or more points).	800
Student council president. Triple participation or honors and awards in selected extracurricular activities (each worth 600 points). Participation in Boys/Girls Nation.	700
High school class president. Editor-in-chief of a school publication. Participation in Boys/Girls State, president of National Honor Society or recipient of a national or state award. Eagle Scout (Boy Scouts) or Gold Award (Girl Scouts). Triple participation or honors and awards in selected extracurricular activities (each worth 500 points).	600
Holder of one or more elective offices in moderately selective organizations. Participation in activities or recipient of awards in moderately selective organizations. Holder of a private pilot's license.	500
Participation in activities or recipient of awards in organizations with limited selectivity.	400
Some participation in organized activities.	300
No participation in organized activities.	200

Faculty Appraisal Score

A candidate's FAS is based on a high school official's evaluation of the candidate's performance. The evaluation is completed by the candidate's English, math, or physics/chemistry instructor. The instructor specifies the degree to which he or she agrees with 12 statements regarding a candidate's demonstrated ability to:

1. show interest and concern for the welfare of others
2. work effectively with others toward group goals
3. influence others in a positive manner
4. communicate effectively in face-to-face discussion
5. communicate effectively in written work
6. set an example of good conduct for others
7. set high standards for own performance in a number of activities
8. maintain composure and perform effectively under pressure
9. adjust to demanding schedule of activities without neglecting school work
10. seek academic challenge beyond that required by normal course work
11. reach sound, logical conclusions based on analysis of facts
12. accept full responsibility for own actions.

In addition, the instructor is asked to write how he or she feels the candidate will perform at the college level in the faculty member's subject-matter area.

Candidate Fitness Assessment

The CFA tests a candidate's strength, agility, speed, and endurance through six tests:

1. basketball throw from kneeling position
2. cadence pull-ups or the flexed-arm hang (women's option)
3. shuttle run
4. modified sit-ups
5. push-ups
6. a one-mile run.

Essays

Candidates are required to write three short essays on the following topics:

1. Explain why you want to attend the United States Military Academy and serve on active duty as an Army officer.
2. What are the most important qualities in becoming a successful USMA cadet and a successful Army officer?
3. West Point and the Army are committed to the idea that respect for others and an understanding of diversity are important leadership traits. Why will you be successful in working with leaders, peers, and subordinates of a gender, color, ethnicity, and/or religion different from your own?

Nomination

Most candidates need a nomination to be considered for admission. Sources of nomination are the vice president; U.S. senators; U.S. representatives; delegates to the House of Representatives from Washington, D.C., Guam, and the Virgin Islands; the governors of American Samoa and Puerto Rico; the resident commissioners of Puerto Rico and the Northern Mariana Islands; or the Secretary of the Army. A limited number of children of members of the armed forces killed or missing in action, or who died of or have a 100-percent service-connected disability, and children of civilian employees who are in missing status may be admitted without a nomination. The president may appoint a limited number of sons and daughters of career military personnel (Army, Navy, Air Force, Marine Corps, and Coast Guard) or Regular Army and Reserve Component members (Army, Army Reserve, and Army National Guard) who meet certain time-in-service or retirement requirements. The president may also appoint children of persons awarded the medal of honor (see U.S. Code, Title 10 for complete details regarding nominations and appointments to USMA).

Who Is Accepted? Who Is Rejected? Who Declines an Acceptance Offer?

As noted above, West Point considers several pieces of information in selecting applicants. In addition, some individuals who receive an offer of admission choose not to attend. Because we obtained complete data on applicants, we were able to examine the characteristics of applicants who were offered admission and who were rejected. Of course, some applicants who were offered admission declined the offer. Table 2.4 compares the admission variable scores of applicants who fall into one of three groups of applicants: rejected, offered admission and accepted, and offered admission but declined. The mean and standard deviation values suggest that although candidates who accepted an offer of admission generally have higher scores than rejected candidates, there is substantial overlap between the WCSs of these two groups. The applicants who declined an admission offer from the USMA have higher application scores than the applicants who accepted, except for

Table 2.4
Average Scores and Standard Deviations of USMA Applicants (Class Years 1992–2001)

	Rejected	Offered/ Admitted	Declined
Number of applicants	9,612	12,334	4,272
WCS	5,399.53 (522.08)	6,012.33 (379.76)	6,183.51 (395.43)
Composite SAT/Converted ACT score	1,154.48 (147.31)	1,268.65 (103.26)	1,293.49 (105.49)
HSRCS	512.40 (111.84)	568.60 (102.96)	614.16 (104.04)
CFA	485.19 (112.53)	557.62 (77.14)	549.18 (75.05)
AAS	537.97 (153.60)	606.65 (110.66)	616.50 (109.35)
EAS	482.35 (131.61)	533.27 (113.91)	539.37 (116.45)
FAS	648.19 (63.94)	674.66 (42.03)	685.81 (38.38)
Gender (% male)	85.2	87.1	84.2

the CFA. But there is significant overlap between the scores of these two groups as the standard deviations suggest.

It is possible for a candidate with a higher WCS to be rejected while a candidate with a lower WCS is offered admission to the USMA. This is due to the fact that the WCS, while a central component of the application, is not the only element used in selecting students. Essays, class composition goals, and other factors that we do not observe in the data influence admission decisions.

Considerations in Using ACT and SAT Scores in the Models

Between 1989 and 1996, both ACT and SAT exams have undergone important changes. The West Point application data we use include applicants for admission for 1988 through 1997 and thus include applicants with different versions of ACT and SAT scores.

In October 1989, ACT content was changed, its scores were recentered, and the test was renamed Enhanced ACT. Enhanced ACT's science reasoning subtest focuses on test-takers' analytical problem-solving skills by utilizing reading material, graphs, and charts, whereas the old ACT's natural sciences subtest tested specific scientific knowledge. Enhanced ACT's reading subtest measures reading ability and comprehension, whereas the old ACT's social studies subtest included questions about U.S. history. Enhanced ACT's math subtest includes trigonometry and pre-algebra and Enhanced ACT's English section focuses on writing skills.

Enhanced ACT results are reported on a 1–36 scale just like the old ACT, but the average composite score increased from 18.6 to 20.6. The ACT technical manual states that results obtained in October 1989 or later are not directly comparable to scores obtained before this date due to differences in the internal structure of the tests and the methodology used for scaling (ACT, 2007). For this reason, we removed all applicants whose old ACT scores were used during calculation of their WCSs.

In April 1995, SAT scores were recentered, resulting in an increase in the mean scores of those who took the test after April 1995. We used SAT conversion tables to recenter SAT verbal and SAT math scores before adding the two components to get the SAT composite score.

ACT scores are calculated by taking the mean of four ACT subtests, before converting the final ACT score to SAT score by a concordance table prepared by ACT and the College Board, the administrators of ACT and SAT tests, respectively.

CHAPTER THREE

Analysis of Success at West Point

The first hurdle a West Point cadet must clear on the way to becoming a commissioned Army officer is to graduate. In the analysis of West Point success, GPAs and orders of merit were unavailable in the data we obtained, so we used graduation as the outcome variable. The graduation variable equals zero if the applicant enrolled but did not graduate from West Point and it equals one if the applicant graduated.[1] West Point graduation rates by year for the sample we used in our analysis are given in Table 3.1. The average graduation rate is 76 percent over the ten-year period, with little difference across graduation years.

We used three logistic regression models and a boosted regression (Schonlau, 2005) to examine if and how the score elements available for use in admissions decisions are related to important outcomes. Because there are other considerations used in the admissions decision for which we do not have information (e.g., class composition goals and written essays), our goal is not to choose among models but to explore the relevance of the score elements for predicting success. The first model we tested uses only the standardized WCSs as predictors. In the second model, we eliminate the WCS but include its standard-

[1] A candidate is considered to be a graduate if his or her application status indicates that they accepted an offer from the USMA and that they received their commission from the USMA. Because of the highly structured environment, cadets are allowed to spend more than four years only under exceptional circumstances, such as illness or injury. Of the 8,606 candidates who graduated in our final sample (see Appendix A for details), 8,316 (96.6 percent) graduated in four years, 211 candidates (2.4 percent) graduated in five years, 69 candidates (0.8 percent) graduated in six years, three candidates graduated in seven years, and seven candidates have missing data on the date of first commission.

Table 3.1
Graduation Rates by Year

USMA Graduation Year of the Application	Graduated	Did Not Graduate	Total	Graduation Rate (%)	Graduation Rate (Analysis Sample[a]) (%)
1992	953	358	1,311	73	75
1993	991	350	1,341	74	76
1994	1,013	309	1,322	77	77
1995	986	262	1,248	79	79
1996	881	308	1,189	74	74
1997	857	354	1,211	71	71
1998	875	272	1,147	76	76
1999	938	249	1,187	79	79
2000	928	258	1,186	78	78
2001	911	281	1,192	76	76
Total	9,333	3,001	12,334	76	76

NOTE: Graduates are not identified directly in the data available to us. As a result, USMA graduation status is inferred from an indication that an individual was accepted and admitted to USMA and was identified by DMDC as being from the USMA accession program.

[a] See Appendix A for details. 1,048 cases were dropped for missing data on the selection component scores and for older SAT or ACT scores for which conversion tables were unavailable. Note that this affected the 1992–1994 classes more than the later classes.

ized components, namely the CEER score, FAS, AAS, EAS, and FAS as predictors. The third model uses the individual components of the WCS as well, but replaces the USMA-calculated CEER score with the composite SAT/converted ACT score calculated by the authors from applicants' SAT and ACT scores and the HSRCS.

Our inclusion of a boosted regression (Model 4) using the same variables as the third model requires a bit more explanation. Boosted logistic regression has the advantage of aggregating the predictions from many weak regression tree classifiers and typically outperforms a single logistic model in predicting outcomes.[2] Boosted regression models can take into account the interactions that are difficult to capture with a simple logistic model (Schonlau, 2005). Our purpose in including a boosted regression model is to obtain an estimate of what would be the best prediction available, given the variables used in the admissions process. Thus, if the boosted regression model is substantially better in predicting outcomes than the simple logistic models, it would suggest that there is a better combination of the available variables that could be discovered for use in the admissions process.

In the end, the point of these variations in models is to assess whether there is a more predictive combination of applicant data than the current WCS.

To facilitate the interpretation of their relationships with the graduation and promotion outcomes, dependent variables were standardized (mean = 0; standard deviation = 1). The logarithm of odds ratios reported in table format in Chapters Three and Four indicate the change in the logarithm of odds of graduation, early promotion to O-4, or promotion to O-5 for a standard deviation change of 1 in the variable. We also report the average marginal effects and standard errors of the coefficients.

---

[2] Each regression tree partitions the dataset according to simple decision rules. The regression tree's prediction for each of the observations in a partition is the average response variable value in that partition (note that the regression tree predicts the errors from the logistic regression). The logistic regression's likelihood function is then updated with the regression tree. A new regression tree is fitted to the residuals from the updated likelihood function and the algorithm continues until a certain number of iterations are complete. The algorithm is run on a randomly selected training part of the dataset, and it is tested on the remaining observations to test the model's adequacy. Besides classification estimates, the boosted regression produces influence values for each variable. The sum of log likelihood increasing across all variables due to a given variable yields the influence of that variable. Influence values are normalized and reported as percentages.

Description of the Sample for the Analysis of the Graduation Outcome

Only applicants with complete application data who accepted an admission offer from West Point were included in the analysis to predict West Point success. Among the candidates who entered West Point, those who are indicated as having received their commission from West Point are assumed to have graduated, and those who did not receive a commission or who received their commission from elsewhere are assumed to have not graduated from West Point. For a detailed description of the sample used in the analysis, please refer to Appendix A.

Restricting the sample to applicants who were admitted to West Point limits our ability to measure the significance and the size of the association between application scores and the graduation outcome. Because it is impossible to observe the USMA graduation outcome for applicants who were not admitted to West Point, the association between application scores and graduation can only be estimated for applicants whose probability of graduation is likely higher compared with the complete pool of applicants, because West Point has already screened and selected those who entered. The result is that the association between the application scores and graduation may be higher for the group of all applicants than is estimated in our models. This problem is known as the "range restriction" problem (See Heckman, 1979; Lawley, 1943; Pearson, 1903; Sackett and Yang, 2000).

Model Results

Table 3.2 summarizes the results from the four models for predicting graduation. In the first three models, WCSs and WCS components are significantly associated with higher probability of graduation. For example, the coefficient for WCSs in Model 1 is 0.263 and significant (see Appendix D for an explanation of how to interpret the model coefficients), indicating that the logarithm of the odds of graduation increases by 0.263 for each 377-point increase in a WCS (corresponding to one standard deviation of the WCS in the sample). Average mar-

Table 3.2
Predicting Graduation from West Point
(Average Marginal Effects in Parentheses)[a] [Standard Errors in Brackets][b]

Scores (Standardized)	Model 1 N: 11,286, Pseudo R^2: 0.0114	Model 2 N: 11,286, Pseudo R^2: 0.0132	Model 3 N: 11,286, Pseudo R^2: 0.0140	Model 4 N: 11,286, Test R^2: 0.0151
WCS	0.263** (0.047) [0.022]	X	X	X
CEER	X	0.189** (0.034) [0.025]	X	X
Composite SAT/ Converted ACT	X	X	0.187** (0.033) [0.024]	27.552
HSRCS	X	X	0.038 (0.007) [0.025]	22.865
CFA	X	0.091** (0.016) [0.023]	0.095** (0.017) [0.023]	15.327
AAS	X	0.066** (0.012) [0.023]	0.072** (0.013) [0.023]	5.122
EAS	X	0.086** (0.015) [0.023]	0.091** (0.016) [0.023]	3.597
FAS	X	0.119** (0.021) [0.023]	0.138** (0.025) [0.023]	25.537

[a] The average marginal effect is the change in the probability of graduation given a change in a score of one standard deviation, averaged over all candidates in the sample. For example, according to the first model, a one standard deviation increase in the WCS increases the probability of graduation by 4.7 percent, on average.

[b] Standard errors of the model coefficients are a measure of the variability of the mean value of coefficients. Because the sample analyzed is just one possible sample of many other samples, the coefficients found by the models might change. Standard errors of the coefficients give a measure of how much variability can be expected given repeated samples from the same population. Smaller standard errors increase our confidence about the precision of model coefficients, because there is less variability around the mean.

* Significant at 0.05 level.

** Significant at 0.01 level.

ginal effects are somewhat more intuitive. In this example, the average marginal effect for a WCS in Model 1 is 0.047. This can be interpreted to mean that one standard deviation increase in a WCS increases the probability of graduation by approximately 4.7 percent.[3]

Our results show that CEER, the components of the CLS (AAS, EAS and FAS), and the CFA are significant predictors of graduation in Model 2. All three components of the WCS—the academic, leadership, and athletic components—are significantly associated with better chances of graduation. In the third model, which breaks the CEER down to its two components, SAT/ACT scores are significantly associated with higher graduation probabilities but the HSRCS is not, contrary to the findings by Niu and Tienda (2009). This suggests that when all the other variables are taken into account (including the SAT/ACT scores), a candidate's HSRCS does not give information on how likely he or she is to graduate in a logistic model. The fourth model we used is a boosted regression model (Schonlau, 2005). The fourth model suggests that the Composite SAT/Converted ACT score, the FAS, and the HSRCS have the biggest influence on the graduation outcome. The influence of the HSRCS in the fourth model might be due to the variable's interaction effects that were not captured in the simple logistic model but that improved the fit of the boosted regression model. As one might expect, these scores also have the largest marginal effect sizes reported for Model 3. An important result from the second and the third models is that both

[3] The average marginal effect can be strictly interpreted as the average rate of change in probability of graduation across all candidates in our sample, given a small amount of change in the standardized WCS. Because the rate of change varies for candidates across the score range (see Appendix D), one standard deviation increase in a WCS does not exactly translate into a 4.7-percent increase in probability. For example, an applicant who scores one standard deviation below the average WCS has a 71.5-percent probability of graduating, whereas an applicant with an average WCS has a 76.6-percent probability of graduating, which is an increase of 5.1 percentage points in the probability of graduation. An applicant who scores one standard deviation above the average on his or her WCS has an 81-percent probability of graduating, or a 4.4-percentage point increase in graduation probability relative to someone at the average WCS. The second increase in the probability of graduation is lower than the first increase because the rate of change in probability of graduation is not constant across the score range.

AASs and CFAs are significant predictors of graduation. The result suggests that physical preparation and academic preparation are associated with increased chances of success. The fourth model reports that the SAT/ACT scores, the FAS, and the HSRCS were the variables that increased the likelihood function the most. In other words, they were more instrumental in predicting the graduation outcome. The fourth model gives little weight to the AAS and the EAS. Note that there is no formal cut-off point for the significance of influence statistics in the fourth model.

Figure 3.1. displays the predicted probability of graduation plotted against the WCS using the results of Model 1. As expected from the model, the curve is negatively accelerating monotonic, indicating

Figure 3.1
Predicted Probability of Graduating from USMA Using Whole Candidate Score

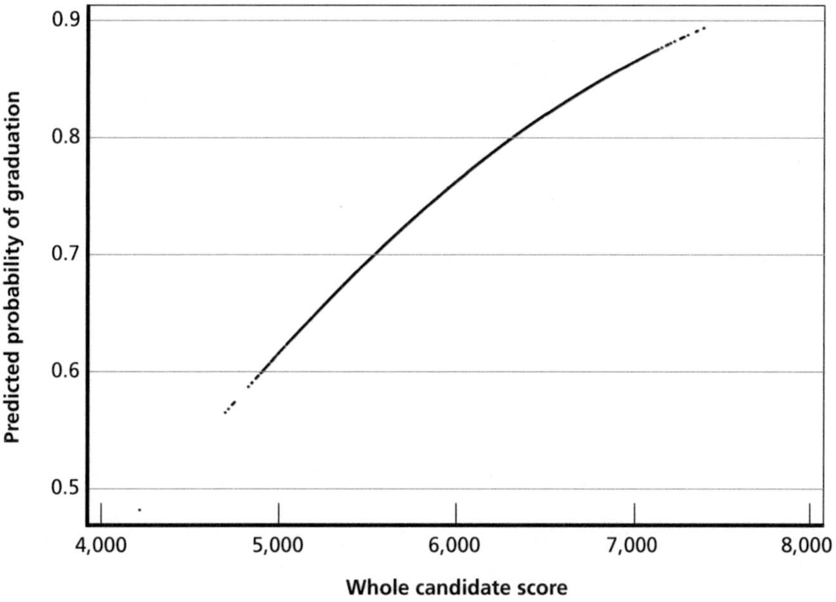

RAND *RR723-3.1*

that increases in WCSs are associated with an increasing probability of graduation.[4]

Although we did not have GPAs available in our data, the validity of the WCS for predicting GPAs and a military performance score (MPS) has been well documented. In a study of a measure of grit in predicting performance at West Point, Duckworth et al. (2007) noted, "Even more striking was the superior prediction by Whole Candidate Score of both MPS ($r = .42$, $p < .001$) and GPA ($r = .64$, $p < .001$)"—both statistically significant. Butler and McCauley (1987) reported correlations of 0.30 and 0.42 between SAT verbal, SAT math, and fourth-year GPAs for the USMA classes of 1982 and 1983. A study by the USMA Office of Institutional Research (Houston, 1970) reported that the correlation of CEER with fourth-year GPAs for the class of 1972 was 0.63. As early as 1948, research into the prediction of GPAs at West Point showed that standardized tests of mathematical concepts, arithmetic reasoning, and language aptitude are significantly related to GPAs (Baier, 1948).

Although our data do not include GPAs, we can be confident from earlier research, as noted above, that the measures of cognitive ability reported here (i.e., CEER and Composite SAT/Converted ACT scores) are significantly related to GPA.

4 We examined the WCS for differential prediction of graduation for race-ethnic and gender groups. For women, as reported by USMA (2012), the WCS was found not to predict graduation. We do not have the data elements required to examine this issue further, but the USMA meeting minutes suggest, as does the RAND NDRI companion report on USAFA admissions (Hardison, Burkhauser, and Hanser, forthcoming), that this is because women leave for nonacademic reasons rather than for academic reasons. Hence, admissions scores that are academically based (i.e., weighing heavily on standardized test scores and high school rank, as WCSs do) would not be expected to predict nonacademic attrition. A comparison of Caucasian males to African-American males, Hispanic males, and Asian males showed no evidence of differential prediction for these racial and ethnic groups.

CHAPTER FOUR

Analysis of Career Success

To measure career success, we used two dichotomous variables. We labeled the first variable "early promotion," which equals one if the candidate was promoted to O-4 within ten years of receiving his or her commission and zero if the candidate reached rank O-4 after ten or more years have passed since first commission date.[1] The second variable is promotion to O-5, which equals one if the candidate was promoted and zero if not. We used the same four models described in the previous chapter but added "dummy" variables for each graduation year in the sample. Graduation year "dummy" variables were added to account for differences in promotion rates that may have varied due to year-to-year changes in promotion opportunity or promotion policies. The 1992 graduation year is excluded and the results from the graduation year variables are interpreted as comparisons with the 1992 baseline year. The coefficients for the graduation years are the log-odds of the outcome variable being equal to one (being promoted to rank O-4 early for the first analysis and being promoted to O-5 in the second analysis) compared with a candidate whose graduation year is 1992. Unlike West Point's relatively stable graduation rates, the graduation year of the application is an important predictor of O-5 promotion, if only because there is a minimum length of service before being considered for O-5 promotion.

---

[1] According to DoD Instruction 1320.13, the desired active-duty list promotion timing to O-4 is ten years (plus or minus one year) and to O-5 is 16 years (plus or minus one year).

Description of the Sample for the Analysis of Early Promotions to O-4

All applicants to USMA with complete application data and who were promoted to O-4 are included in the second part of the analysis. Thus, some individuals included for this analysis may have been commissioned through ROTC or OCS. Applicants who reached O-4 in fewer than seven years and those who reached O-5 in fewer than ten years are excluded from the analysis as unidentified data errors or unlikely outliers. Table 4.1 shows the number of individuals in the sample and their promotion outcomes to O-4.

Model Results

Table 4.2 summarizes the model results, which are not especially profound. Among the components of the WCS across all models, only HSRCS in Model 3 is significantly associated with higher odds of being promoted early to O-4. Haggerty (1963, p. 11) also reported that neither academic course grades nor overall class standing at West

Table 4.1
Early Promotion to O-4 (Less Than Ten Years)

USMA Graduation Year of the Application	Not Early Promotion	Early Promotion	Total	Early Promotion Percentage
1992	658	27	685	4
1993	598	37	635	6
1994	688	54	742	7
1995	793	76	869	9
1996	790	106	896	12
1997	686	219	905	24
1998	487	339	826	41
1999	633	157	790	20
2000	423	362	785	46
2001	354	349	703	50
Total	6,110	1,726	7,836	22

Table 4.2
Predicting Early Promotion to O-4
(Average Marginal Effects in Parentheses) [Standard Errors in Brackets]

Scores (Standardized)	Model 1 N: 7,836, Pseudo R^2: 0.1494	Model 2 N: 7,836, Pseudo R^2: 0.1497	Model 3 N: 7,836, Pseudo R^2: 0.1505	Model 4 N: 7,836, Test R^2: 0.1339
WCS	0.029 (0.004) [0.030]	X	X	X
CEER	X	0.004 (0.001) [0.033]	X	X
Composite SAT/ converted ACT	X	X	–0.040 (–0.006) [0.005]	8.957
HSRCS	X	X	0.089** (0.013) [0.005]	5.060
CFA	X	0.008 (0.001) [0.032]	0.008 (0.001) [0.005]	11.242
AAS	X	0.035 (0.005) [0.033]	0.033 (0.005) [0.005]	1.570
EAS	X	0.031 (0.004) [0.031]	0.023 (0.003) [0.004]	2.146
FAS	X	0.019 (0.003) [0.035]	–0.014 (–0.002) [0.005]	11.087
Graduation year 1993	0.408 (0.059) [0.259]	0.413 (0.060) [0.259]	0.421 (0.061) [0.038]	1.002
Graduation year 1994	0.648** (0.094) [0.242]	0.645** 0.093 [0.242]	0.651** (0.094) [0.035]	0.549
Graduation year 1995	0.848** (0.123) [0.230]	0.840** 0.122 [0.230]	0.851** (0.123) [0.033]	0.165
Graduation year 1996	1.183** (0.172) [0.222]	1.176** 0.170 [0.222]	1.193** (0.173) [0.032]	0.111
Graduation year 1997	2.052** (0.297) [0.211]	2.042** 0.296 [0.212]	2.071** (0.300) [0.030]	3.497

Table 4.2—Cont.

Scores (Standardized)	Model 1 N: 7,836, Pseudo R^2: 0.1494	Model 2 N: 7,836, Pseudo R^2: 0.1497	Model 3 N: 7,836, Pseudo R^2: 0.1505	Model 4 N: 7,836, Test R^2: 0.1339
Graduation year 1998	2.829** (0.410) [0.209]	2.820** (0.409) [0.209]	2.843** (0.411) [0.030]	17.419
Graduation year 1999	1.798** (0.261) [0.216]	1.792** (0.260) [0.216]	1.804** (0.261) [0.031]	1.615
Graduation year 2000	3.039** (0.441) [0.209]	3.030** (0.439) [0.210]	3.058** (0.443) [0.030]	16.979
Graduation year 2001	3.177** (0.461) [0.210]	3.154** (0.457) [0.214]	3.215** (0.465) [0.030]	18.602

* Significant at 0.05 level.

** Significant at 0.01 level.

Point were predictive of promotion to a temporary rank of major (O-4) within six years of graduation.[2] A WCS is not associated with higher or lower probability of early promotion to O-4.

Description of the Sample for the Analysis of Promotions to O-5

Only applicants with complete application data and who were promoted to O-4 are included in the analysis of promotion to O-5 (Table 4.3). Applicants who reached O-4 in fewer than seven years and those who reached O-5 in fewer than ten years are excluded from the analysis as before. As of 2012, there were no promotions to O-5 for graduation years later than 1999 and only seven promotions to O-5 for graduation year 1999. Gradu-

---

[2] Haggerty (1963) reports this result for the USMA class of 1953. Times to promotion have differed over the past half-century and her report of promotion to "temporary rank of major" suggests that it was early promotion to major that served as the outcome variable in her report.

Table 4.3
Promotion to O-5 by Year

USMA Graduation Year of the Application	Promoted to O-5	Not Promoted to O-5	Total	O-5 Promotion Percentage
1992	493	192	685	72
1993	463	172	635	73
1994	512	230	742	69
1995	589	280	869	68
1996	439	457	896	49
1997	149	756	905	16
1998	32	794	826	4
1999	7	783	790	1
Total	2,684	3,664	6,348	42

ation years 2000 and 2001 are excluded from the analysis for promotion to O-5 because not enough time had elapsed for promotions to O-5 to occur.

Model Results

Model results for the O-5 promotion sample are shown in Table 4.4. In the first model, a higher WCS is associated with higher odds of promotion to O-5. Application years 1996, 1997, 1998, and 1999 are associated with lower odds of promotion compared with application year 1992. In the second model, higher CEERs, CFAs, EASs, and FASs are associated with better odds of promotion to O-5. This is somewhat consistent with Bartone, Snook, and Tremble (2002), who reported that CEER was significantly related to military development grades for upperclassmen at West Point. They also note, "The college entrance exam scores, a measure of general intellectual ability assessed as part of the college application process, is a consistent predictor of later leader development scores for West Point cadets" (p. 330).

Candidates whose graduation years are 1995, 1996, 1997, 1998, and 1999 have decreasing odds of promotion to O-5 compared with candidates who graduated in 1992. AASs are not associated with better odds of promotion to O-5. In the third model, the composite SAT/con-

Table 4.4
Predicting Promotion to O-5
(Average Marginal Effects in Parentheses) [Standard Errors in Brackets]

Scores (Standardized)	Model 1 N: 6,348, Pseudo R^2: 0.3172	Model 2 N: 6,348, Pseudo R^2: 0.3178	Model 3 N: 6,348, Pseudo R^2: 0.3182	Model 4 N: 6,348, Test R^2: 0.3202
WCS	0.248** (0.038) [0.032]	X	X	X
(CEER)	X	0.194** (0.030) [0.034]	X	X
Composite SAT/ converted ACT	X	X	0.133** (0.021) [0.035]	3.311
HSRCS	X	X	0.126** (0.019) [0.037]	4.329
CFA	X	0.067* (0.010) [0.033]	0.073* (0.011) [0.034]	7.741
AAS	X	0.009 (0.001) [0.023]	0.012 (0.002) [0.033]	1.311
EAS	X	0.065* (0.010) [0.023]	0.061 (0.010) [0.033]	1.591
FAS	X	0.083* (0.013) [0.034]	0.077* (0.012) [0.035]	4.555
Graduation year 1993	0.023 (0.004) [0.124]	0.029 (0.004) [0.124]	0.032 (0.005) [0.124]	1.713
Graduation year 1994	–0.148 (–0.023) [0.117]	–0.163 (–0.025) [0.117]	–0.159 (–0.025) [0.117]	1.281
Graduation year 1995	–0.206 (–0.032) [0.113]	–0.227* (–0.035) [0.113]	–0.217 (–0.034) [0.113]	1.452
Graduation year 1996	–1.012** (–0.157) [0.109]	–1.032** (–0.160) [0.110]	–1.012** (–0.157) [0.110]	2.328
Graduation year 1997	–2.602** (–0.404) [0.125]	–2.627** (–0.407) [0.126]	–2.617** (–0.406) [0.126]	26.439

Table 4.4—Cont.

Scores (Standardized)	Model 1 N: 6,348, Pseudo R^2: 0.3172	Model 2 N: 6,348, Pseudo R^2: 0.3178	Model 3 N: 6,348, Pseudo R^2: 0.3182	Model 4 N: 6,348, Test R^2: 0.3202
Graduation year 1998	−4.214** (−0.654) [0.200]	−4.232** (−0.656) [0.201]	−4.223** (−0.654) [0.201]	22.953
Graduation year 1999	−5.712** (−0.886) [0.389]	−5.726** (−0.888) [0.390]	−5.666** (−0.878) [0.390]	20.997

* Significant at 0.05 level.

** Significant at 0.01 level.

verted ACT score, HCRCS, CFA, and FAS are associated with higher odds of promotion to O-5. Note that both the second and third models suggest that the CFA is associated with higher chances of promotion to O-5 as was found in the graduation analysis, but the relationship does not hold for the AAS. Candidates whose graduation years are 1996, 1997, 1998, and 1999 have decreasing odds of promotion to O-5 compared with candidates who graduated in 1992.

The boosted regression model suggests that the year variables 1997, 1998, and 1999 have the greatest influence on the promotion odds. Models did not perform better than the selection process for graduation year 1992—as shown in Appendix B, which includes a comparison of the four models in selecting the cohort with the highest O-5 promotion percentage.

Figure 4.1. displays the predicted probability of promotion to O-5, given promotion to O-4, for the graduating classes of 1992 through 1996 plotted against the WCS. The curve is very slightly negatively accelerating monotonic, indicating that increases in WCSs are associated with an increasing probability of promotion to O-5.[3]

[3] WCS prediction of promotion to O-5 does not differ for any of the gender, racial, or ethnic groups.

Figure 4.1
Predicted Probability of Promotion to O-5 USMA Using Whole Candidate Score (Graduation Years 1992–1996)

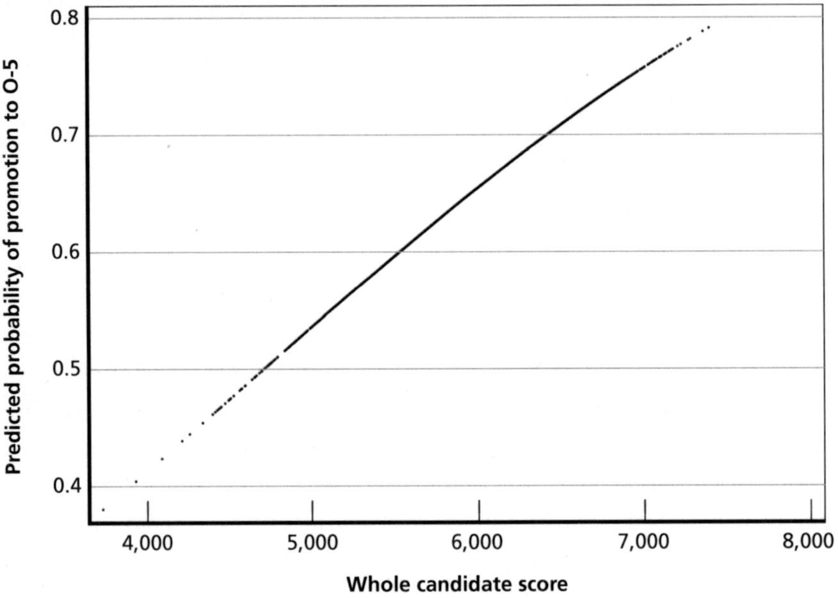

RAND *RR723-4.1*

CHAPTER FIVE

Conclusions

Admission to West Point is a highly complex process. We observe that an applicant's WCS, while important, is not the sole determinant of admission, even if the applicant is deemed academically, physically, and medically qualified. Many applicants who were not offered admission had higher WCSs than applicants who were offered admission. Whether this results from some combination of judgments of the essay component of applications that were not available to us and/or class composition goals is unknown.

Nonetheless, results indicate that the WCS has a statistically significant relationship with both graduation from the USMA and probability of promotion to O-5 in the Army. All subelements of the WCS except HSRCS make significant contributions to the prediction of graduation, but HSRCS is significant in the prediction of both early O-4 promotion and O-5 promotion. Both the CFA and the AAS are statistically significant predictors of graduation. CFAs are also significantly associated with higher probability of promotion to O-5. Academics and fitness are both important criteria in the selection of candidates. Table 5.1 summarizes the significance of different application scores in explaining graduation and promotion to O-5 outcomes. Variables that are significant are in gray cells. Early promotion to O-4 is excluded because none of the variables were significant.

However, the magnitude of the relationships we observe are likely underestimates due to selection effects—the graduation and promotion outcomes for applicants who were not admitted to USMA, to another service academy, or to an ROTC program (in the case of pro-

Table 5.1
Significant Variables in the Logistic Models (Significant Variables Have Gray Backgrounds)

Graduation Model 1	Graduation Model 2	Graduation Model 3	Promotion to O-5 Model 1	Promotion to O-5 Model 2	Promotion to O-5 Model 3
WCS	X	X	WCS	X	X
	CEER	X		CEER	X
	X	Composite SAT/ converted ACT		X	Composite SAT/ converted ACT
	X	HSRCS		X	HSRCS
	CFA	CFA		CFA	CFA
	AAS	AAS		AAS	AAS
	EAS	EAS		EAS	EAS
	FAS	FAS		FAS	FAS

motion outcomes) cannot be observed. Our regression models based on the component scores that make up the WCS did not indicate that a different weighting or combination of components could improve the selection process.[1]

This study shows that applicants scoring higher on WCSs have a greater probability of graduation, the first necessary step for entering a career as an Army officer, and of promotion to O-5, traditionally considered the mark of achievement of a successful military career. In our judgment, the use of WCSs as the primary basis for admission decisions at West Point is warranted.

[1] Appendixes B and E explore how predictions could be improved with the weights from our models, and indicate that no substantial improvements were found.

APPENDIX A

Sample Used in Explaining West Point Success, Early Promotion, and O-5 Promotion

Table A.1
Sample for West Point Success Analysis

	Dropped	Remaining Observations
The original dataset contains all applications between 1992 and 2001.		130,265
Keep only one application per applicant by retaining only the latest one.	5,364	124,901
Drop all observations with an application result that is not "Admitted." Graduation is only analyzed for applicants who were accepted to West Point and who accepted the offer.	112,567	12,334
Drop all observations with old ACT scores.	1,018	11,316
Drop all observations without an FAS.	15	11,301
Drop all observations without a WCS.	1	11,300
Drop all observations without a CFA.	4	11,296
Drop all observations without an HSRCS.	2	11,294
Drop all observations from which SAT scores were used but SAT verbal and SAT math scores are missing or one of these scores is lower than 200.	1	11,293
Drop all observations for which ACT scores were used but one of ACT English, math, science reasoning, or reading scores is missing.	7	11,286

Table A.2
Sample for Career Success Analysis, Early Promotion to O-4

	Dropped	Remaining Observations
The original dataset contains all applications between 1992 and 2001.		130,265
Keep only one application per applicant by retaining only the latest one.	5,364	124,901
Drop an application if it was not accepted and the applicant was later commissioned from West Point. These applicants probably applied to West Point again and got accepted, but the application that was accepted by West Point is not included in the dataset.	252	124,649
Drop all observations without a WCS.	25,964	98,685
Drop all observations without a CFA.	57,555	41,130
Drop all observations without an HSRCS.	5	41,125
Drop all observations without an FAS.	4,227	36,898
Drop all observations from which SAT scores were used but SAT verbal and SAT math scores are missing or one of these scores is lower than 200.	69	36,829
Drop all observations for which ACT scores were used but one of ACT English, math, science reasoning, or reading scores is missing.	207	36,622
Drop all observations without a date that indicates the year the applicant received a commission.	18,193	18,429
Drop all observations for which the year of receiving a commission is before the applicant's graduation year.	499	17,930
Drop all observations for which the year of receiving a commission is at least three years later than the applicant's graduation year.	878	17,052
Drop all observations for which the number of years until promotion to O-5 is smaller than ten years, which is the minimum amount of time to be promoted to O-5 after receiving a commission.	3	17,049
Drop all observations belonging to applicants who were not promoted to O-4.	8,434	8,615
Drop all observations with missing gender values.	4	8,611
Drop all observations with ACT composite score lower than 11; there is no SAT conversion value for ACT values smaller than 11.	1	8,610
Drop all observations with old ACT scores.	774	7,836

Table A.3
Sample for Career Success Analysis, Promotion to O-5

	Dropped	Remaining Observations
Original dataset containing all applications between 1992 and 2001.		130,265
Keep only one application per applicant by retaining only the latest one.	5,364	124,901
Drop an application if it was not accepted and the applicant was later commissioned from West Point. These applicants probably applied to West Point again and got accepted, but the application that was accepted by West Point is not included in the dataset.	252	124,649
Drop all observations without a WCS.	25,964	98,685
Drop all observations without a CFA.	57,555	41,130
Drop all observations without an HSRCS.	5	41,125
Drop all observations without an FAS.	4,227	36,898
Drop all observations from which SAT scores were used but SAT verbal and SAT math scores are missing or one of these scores is lower than 200.	69	36,829
Drop all observations for which ACT scores were used but one of ACT English, math, science reasoning, or reading scores is missing.	207	36,622
Drop all observations without a date that indicates the year the applicant received a commission.	18,193	18,429
Drop all observations for which the year of receiving a commission is before the applicant's graduation year.	499	17,930
Drop all observations for which the year of receiving a commission is at least three years later than the applicant's graduation year.	878	17,052
Drop all observations for which the number of years until promotion to O-5 is smaller than ten years, which is the minimum amount of time to be promoted to O-5 after receiving a commission.	3	17,049
Drop all observations belonging to applicants who were not promoted O-4.	8,434	8,615
Drop all observations with missing gender values.	4	8,611
Drop all observations with ACT composite score lower than 11; there is no SAT conversion value for ACT values smaller than 11.	1	8,610
Drop all observations with old ACT scores.	774	7,836
Drop all observations with graduation year 2000 and 2001. There are no candidates who were promoted to O-5.	1,488	6,348

NOTE: The final sample size of 6,348 represents all applicants for whom we have complete applicant data and who were promoted to O-4. Some of these applicants were not offered admission or did not attend USMA.

APPENDIX B

Model Comparison for Predicting Promotion to O-5

To assess the performance of the models we used to predict promotion to O-5, we looked at promotion outcomes of different cohorts we selected with the models. For each candidate in the promotion to O-5 sample from graduation year 1994, we found the probability of being promoted to grade O-5 according to the three logistic models and the boosted regression model fitted to data from 1992 and 1993 cohorts. We excluded later years from this analysis because the promotion to O-5 rates are significantly different from the first three years' cohorts. Each candidate received four different probabilities of promotion to O-5 according to four different models. For each model, we ranked the observations according to their estimated O-5 promotion probability. In the data, we observe that 476 candidates were offered admission to West Point out of the 712 candidates from the 1994 cohort that remained in our sample after removing missing values and irregular cases (described in Appendix A), although 99 of these declined admission. Therefore, we counted how many of the first 476 observations were actually promoted to O-5 for each of the four rankings. The models did not completely agree with the West Point admission offer that was observed for graduation year 1994, but the overall rate of O-5 promotion did not change significantly. The results, summarized in Tables B1–B5, suggest that in terms of increasing the percentage of O-5 promotion among candidates who were offered West Point admission, the models did not perform better than the selection methodology used during the selection of the graduation cohort 1994. The USMA selection's promotion percentage is 72 percent and candidates selected by the models have a promotion to O-5 rate of 71 percent. In all the models and in the USMA selection, selected candidates outperform nonselected candidates in terms of rates of promotion to grade O-5.

Table B.1
USMA Selection and O-5 Promotion for the 1994 Cohort

	Promoted to O-5			
Selected by USMA	Yes	No	**Total**	**Promotion Percentage**
Yes	342	134	476	72
No	170	96	266	64
Total	512	230	742	

Table B.2
Prediction Performance of Model 1 for the 1994 Cohort

	Promoted to O-5			
Selected by Model 1	Yes	No	**Total**	**Promotion Percentage**
Yes	336	140	476	71
No	176	90	266	66
Total	512	230	742	

Table B.3
Prediction Performance of Model 2 for the 1994 Cohort

	Promoted to O-5			
Selected by Model 2	Yes	No	**Total**	**Promotion Percentage**
Yes	336	140	476	71
No	176	90	266	66
Total	512	230	742	

Table B.4
Prediction Performance of Model 3 for the 1994 Cohort

	Promoted to O-5			
Selected by Model 3	Yes	No	**Total**	**Promotion Percentage**
Yes	338	138	476	71
No	174	92	266	65
Total	512	230	742	

Table B.5
Prediction Performance of Model 4 for the 1994 Cohort

	Promoted to O-5			
Selected by Model 4	Yes	No	**Total**	**Promotion Percentage**
Yes	338	138	476	71
No	174	92	266	65
Total	512	230	742	

APPENDIX C

Correlations Among the Analysis Variables

Table C.1
Correlation Matrix for Variables in the Graduation Sample (n = 11,286)

	WCS	Converted ACT/SAT	CEER	HSRC	EAS	AAS	FAS	CFA
WCS	1.00							
Composite SAT/ Converted ACT	0.62 0.00							
CEER	0.86 0.00	0.78 0.00	1.00					
HSRC	0.71 0.00	0.25 0.00	0.75 0.00	1.00				
EAS	0.44 0.00	0.08 0.00	0.16 0.00	0.21 0.00	1.00			
AAS	0.06 0.00	−0.29 0.00	−0.27 0.00	−0.13 0.00	−0.06 0.00	1.00		
FAS	0.41 0.00	0.08 0.00	0.27 0.00	0.37 0.00	0.20 0.00	0.02 0.01	1.00	
CFA	0.08 0.000	−0.20 0.000	−0.18 0.000	−0.09 0.000	−0.05 0.000	0.20 0.000	0.01 0.562	1.00

Table C.2
Correlation Matrix for Variables in the Promotion Sample (n = 6,348)

	WCS	Converted ACT/SAT	CEER	HSRC	EAS	AAS	FAS	CFA
WCS	1.00							
Composite SAT/ Converted ACT	0.64 0.00	1.00						
CEER	0.86 0.00	0.80 0.00	1.00					
HSRC	0.70 0.00	0.30 0.00	0.76 0.00	1.00				
EAS	0.40 0.00	0.05 0.00	0.13 0.00	0.18 0.00	1.00			
AAS	0.24 0.00	−0.17 0.00	−0.13 0.00	−0.06 0.00	0.03 0.01	1.00		
FAS	0.44 0.00	0.12 0.00	0.31 0.00	0.41 0.00	0.20 0.00	0.06 0.00	1.00	
CFA	0.21 0.00	−0.10 0.00	−0.10 0.00	−0.07 0.00	0.00 0.87	0.30 0.00	0.05 0.00	1.00

Table C.3
Correlation Matrix for Variables in the Sample with No Missing Values (n = 33,463)

	WCS	Converted ACT/SAT	CEER	HSRC	EAS	AAS	FAS	CFA
WCS	1.00							
Composite SAT/ Converted ACT	0.70 0.00	1.00						
CEER	0.88 0.00	0.84 0.00	1.00					
HSRC	0.69 0.00	0.33 0.00	0.74 0.00	1.00				
EAS	0.45 0.00	0.14 0.00	0.21 0.00	0.25 0.00	1.00			
AAS	0.27 0.00	−0.11 0.00	−0.09 0.00	−0.05 0.00	0.03 0.00	1.00		
FAS	0.46 0.00	0.17 0.00	0.33 0.00	0.40 0.00	0.22 0.00	0.09 0.00	1.00	
CFA	0.27 0.00	−0.06 0.00	−0.04 0.00	−0.02 0.00	0.02 0.00	0.35 0.00	0.07 0.00	1.00

APPENDIX D

Interpreting Coefficients from the Logistic Regression

A probability can be calculated by dividing the number of times an event of interest occurs by the number of all observed events. For example, the probability of graduation from USMA for the graduation year 1992 is calculated by dividing the number of students in the 1992 cohort who graduated (953) by the number of all students in the same cohort (1,311). The probability of a student in the 1992 cohort to graduate is thus 73 percent, disregarding the differences in other characteristics. The odds of an event occurring are calculated by dividing the number of times the event of interest occurs by the number of times the event of interest did not occur. In this case, the odds of graduation can be found by dividing the number of graduations (953) by the number of students who did not graduate (358), making the odds ratio of graduation 2.67. The odds ratio between expected graduation outcomes of two candidates is found by dividing the odds of graduation of one candidate with the odds of graduation of the other candidate.

The coefficients in the logistic regression describe the change in the logarithm of the odds ratio between two candidates when one candidate's score is increased by one standard deviation. For instance, the coefficient for the standardized WCS is 0.263 in Table 3.2 in Chapter Three. This means that increasing the WCS by one standard deviation increases the logarithm of the odds ratio by 0.263. If we raise the coefficient value to the power of e (2.714...) we obtain the odds ratio. In this case, the odds ratio is 1.3. The interpretation is that a candidate with a WCS that is one standard deviation higher than another candidate has 1.3 times the odds of graduation of the candidate with the lower WCS. A candidate with a standardized WCS of 0 has a probability of graduation of 0.766, probability of failure of 0.234, and odds of graduation of 3.269:

$$P(Graduate|stdwcs = 0) = 0.766 = \frac{e^{1.184+0.263\times 0}}{1 + e^{1.184+0.263\times 0}}$$

$$P(Not\ Graduate|stdwcs = 0) = 1 - 0.766 = 0.234$$

$$Odds\ of\ graduation\ for\ a\ candidate\ with\ stdwcs\ of\ 0 = \frac{0.766}{0.234} = 3.269$$

A candidate with a standardized WCS of 1 has a probability of graduation of 0.81, probability of failure of 0.19, and odds of graduation of 4.255:

$$P(Graduate|stdwcs = 1) = 0.810 = \frac{e^{1.184+0.263\times 1}}{1 + e^{1.184+0.263\times 1}}$$

$$P(Not\ Graduate|stdwcs = 1) = 1 - 0.81 = 0.19$$

$$Odds\ of\ graduation\ for\ a\ candidate\ with\ stdwcs\ of\ 1 = \frac{0.81}{0.19} = 4.255$$

Therefore, an applicant with a WCS one standard deviation above the average has an increase of .044 in probability of graduating over an applicant with the average WCS.

The odds ratio of graduation for the standardized WCS is:

$$\frac{4.255}{3.269} = 1.3.$$

The logarithm of 1.3 is 0.263, the value given in Table 3.2 in Chapter Three. The odds ratio remains the same at all levels of a score. The odds of graduation for a candidate with a standardized whole candidate score of 2 is calculated by first calculating the probability of graduation:

$$P(Graduate|stdwcs = 2) = 0.847 = \frac{e^{1.184+0.263\times 2}}{1 + e^{1.184+0.263\times 2}}$$

$$P(Not\ Graduate|stdwcs = 2) = 1 - 0.847 = 0.153$$

$$Odds\ of\ graduation\ for\ a\ candidate\ with\ stdwcs\ of\ 2 = \frac{0.847}{0.153} = 5.538$$

The odds ratio between a candidate with a standardized WCS of 2 and a candidate with a score of 1 is the same as the difference if the candidates had scores of 1 and 0:

$$\frac{5.538}{4.255} = 1.3.$$

APPENDIX E

Comparison of the Receiver Operating Characteristic Curves

The boosted regression model has the best explanatory power among all models. It has the highest area under the Receiver Operating Characteristic (ROC) curve. The area under the ROC curve tells how well the model correctly identifies who will graduate and who will not.

A trade-off common to all tests is that the better a test can identify positives in a sample as positives (its sensitivity), the more negatives it also identifies as positives (false positives). A test's sensitivity can be adjusted by choosing different "cutoff" points, above which the test identifies an observation as positive. A higher cutoff point means it is less likely for a candidate identified as positive to actually be negative (lower false positive rate), because the standard for being identified as positive is higher. But a higher cutoff point also means that more of the candidates identified as negative will actually be positive (higher false negative rate). Similarly, a lower cutoff point means that fewer of the candidates identified as negative are actually positives (lower false negative rate), but it also means more of the candidates identified as positive will actually be negative (higher false positive rate).

The ROC curve plots sensitivity and false positive rates from all cutoff points as an indicator of the accuracy of the test. Sensitivity is defined as:

$$\frac{True\ Positive}{True\ Positive + False\ Negative} = sensitivity$$

where "true positive" is the number of candidates the model predicts will graduate and who graduated, and "false negative" is the number of candidates the model predicts will not graduate but who graduate.

False positive rate is defined as:

$$\frac{False\ Positive}{se\ Positive + True\ Negative} = false\ positive\ r$$

where "false positive" is the number of candidates the model predicts will graduate but who did not graduate, and "true negative" is the number of candidates the model predicts will not graduate and who did not graduate.

The area under the ROC curve is 1 for a perfect test. All candidates identified by the test as positives are indeed positives and all negatives identified by the test are really negatives in a perfect test. The area under the ROC curve is 0.5 for a useless test, because the test does not give any information on the candidate.

Table E.1 summarizes the areas under the ROC curves for different models used in predicting graduation.

Only the boosted regression model has an area under the ROC curve that is statistically significantly higher than the first model (0.607 is significantly higher than 0.576).

Table E.2 summarizes the areas under the ROC curves for different models used to predict promotion to O-5.

Table E.1
ROC Curve Results for Models Predicting Graduation

Model	Observation	ROC Area	Standard Error	95% Confidence Interval	Test of Difference from First Model
1 (WCS only model)	11,286	0.576	0.007	0.563–0.589	
2	11,286	0.581	0.006	0.569–0.593	Prob > chi^2 = 0.095
3	11,286	0.582	0.006	0.569–0.594	Prob > chi^2 = 0.119
4 (boosted model)	11,286	0.607	0.006	0.594–0.619	Prob > chi^2 = 0

Table E.2
ROC Curve Results for Models Predicting Promotion to O-5

Model	Observation	ROC Area	Standard Error	95% Confidence Interval	Test of Difference from First Model
1 (WCS only model)	6,348	0.837	0.005	0.827–0.846	
2	6,348	0.837	0.005	0.827–0.847	Prob > chi^2 = 0.6856
3	6,348	0.837	0.005	0.827–0.847	Prob > chi^2 = 0.5804
4 (boosted model)	6,348	0.852	0.0047	0.842–0.861	Prob > chi^2 = 0

Only the boosted regression model has an area under the ROC curve that is statistically significantly higher than the first model.

References

ACT, "The ACT Technical Manual," 2007, p. 38.

Baier, D. E., "Selection and Evaluation of West Point Cadets," *Educational and Psychological Measurement*, Vol. 8, No. 2, 1948, pp. 93–199.

Bartone, P. T., S. A. Snook, and T. R. Tremble, "Cognitive and Personality Predictors of Leader Performance in West Point Cadets," *Military Psychology*, Vol. 14, No. 4, 2002, pp. 321–338.

Butler, R. P., and C. McCauley, "Extraordinary Stability and Ordinary Predictability of Academic Success at the United States Military Academy," *Journal of Educational Psychology*, Vol. 79, No.1, 1987, pp. 83–86.

DeAngelo, L., R. Franke, S. Hurtado, J. H. Pryor, and S. Tran, *Completing College: Assessing Graduation Rates at Four-Year Institutions*, Los Angeles: Higher Education Research Institute, UCLA, 2011.

DoD Instruction 1320.13, *Commissioned Officer Promotion Reports*, Washington, D.C.: U.S. Department of Defense, July 22, 2009.

Duckworth, A. L., C. Peterson, M. D. Matthews, and D. R. Kelly, "Grit: Perseverance and Passion for Long-Term Goals," *Journal of Personality and Social Psychology*, Vol. 92, No. 6, 2007, pp. 1087–1101.

Geiser, Saul, and Roger Studley, "UC and the SAT: Predictive Validity and Differential Impact of the SAT I and SAT II at the University of California," *Educational Assessment*, Vol. 8, No. 1, 2002, pp. 1–26.

Geiser, Saul, and V. Maria Santelices, *Validity of High-School Grades in Predicting Student Success Beyond the Freshman Year: High-School Record vs. Standardized Tests as Indicators of Four-Year College Outcomes*, Berkeley, Calif.: University of California, Berkeley, Center for Studies in Higher Education, Research & Occasional Paper Series, CSHE.6.07, 2007.

GAO—*See* General Accounting Office.

General Accounting Office, *DOD Service Academies, Problems Limit Feasibility of Graduates Directly Entering the Reserves*, Washington, D.C., March 1997.

Haggerty, H., *Status Report on Research for the U.S. Military Academy (Cadet Leaders Task)*, Washington, D.C.: U.S. Army Personnel Research Office, Technical Research Report 1133 (DTIC Accession Number AD0432090), 1963.

Hardison, Chaitra M., Susan V. Burkhauser, and Lawrence M. Hanser, *United States Service Academy Admissions: Volume 2, Selecting for Success at the United States Air Force Academy and As an Officer*, Santa Monica, Calif.: RAND Corporation, RR-744-OSD, forthcoming.

Heckman, James J., "Sample Selection Bias as a Specification Error," *Econometrica*, Vol. 47, 1979, pp. 153–161.

Houston, John W., *Background and Predicted Success of Cadets in the Class of 1973 with Comparisons with Previous Classes*, West Point, N.Y.: Office of Institutional Research, Report No. XB1-01-70-007, ERIC Document Reproduction Service No. ED111259, 1970.

Lawley, D. N., "A Note on Karl Pearson's Selection Formulae," *Proceedings of the Royal Society of Edinburgh, LXII-Part I (I)*, 1943, pp. 28–30.

McDonald, Deborah J., "Director's Philosophy—Values, Vision, Mission, Message," memorandum for all admissions department personnel, West Point, N.Y., October 1, 2008.

Niu, X. Sunny, and Marta Tienda, "Testing, Ranking and College Performance: Does High School Matter?" *Rassegna Italiana di Sociologia*, Vol. 2, April 1, 2012, pp. 199–226.

Pearson, Karl, "Mathematical Contributions to the Theory of Evolution—XI. On the Influence of Natural Selection on the Variability and Correlation of Organs," *Philosophical Transactions*, Vol. 200, No. 321-330, 1903, pp. 1–66.

Sackett, Paul R., and Hyuckseung Yang, "Correction for Range Restriction: An Expanded Typology," *Journal of Applied Psychology*, Vol. 85, 2000, pp. 112–118.

Schonlau, Matthias, "Boosted Regression (Boosting): An Introductory Tutorial and a Stata Plugin," *Stata Journal*, Vol. 5, 2005, pp. 330–354.

Smithsonian Institution, "West Point in the Making of America," *National Museum of American History* website, undated. As of December 27, 2013: http://americanhistory.si.edu/westpoint/history_1a.html

United States Military Academy at West Point, "2011–2012 West Point Catalog," web page, undated-a. As of December 27, 2013 (page no longer accessible): http://www.usma.edu/admissions/shared%20documents/catalog_2011-12.pdf

———, "About the United States Military Academy Preparatory School," web page, undated-b. As of June 16, 2014: http://www.usma.edu/usmaps/SitePages/About.aspx

———, Minutes, Board of Visitors Organizational Meeting, West Point, N.Y.: June 14, 2012. As of May 27, 2014:
http://www.usma.edu/bov/siteassets/meeting%20minutes/Signed%20Meeting%20Minutes%2020120614.pdf

U.S. Code, Title 10, Subtitle B, Part III, Chapter 403, § 4342: Cadets: appointment; numbers, territorial distribution.

U.S. Department of Defense, *Career Progression of Minority and Women Officers*, Washington, D.C.: Office of the Undersecretary of Defense (Personnel and Readiness), 1999.

USMA—*See* United States Military Academy at West Point.

White House, "Steps for the United States Military Academies Application Process," web page, undated. As of December 27, 2013:
http://www.whitehouse.gov/administration/vice-president-biden/academy-nominations/steps